I Can Draw

COLORING & ACTIVITY BOOK

SOCIAL-EMOTIONAL SUPPORT

AHotMessDotCom

FACIAL EXPRESSIONS MADE FROM SHAPES

Illustrated & Designed by Takora Lineberger of AHotMessDotCom LLC

June 2023
Book design by Takora Lineberger
Written by Takora Lineberger
Illustrated by Takora Lineberger
ISBN 979-8-9892871-1-6
Published by AHotMessDotEDU
www.AHotMessDotEDU.com
www.AHotMessDotCom.com

<u>Dedications</u>

This book is dedicated to all of the parents and children who take pride in learning from home in addition to school.

A special dedication to Taren, my one and only child. This book is for us!

I hope this book encourages my sisters to follow through with their dreams and desires. We're UP from here.

:)

ABOUT THIS BOOK

This book was created for learners to have fun alongside parents and teachers. Time spent together is the most valuable time there is.

Social Emotional Learning

To aid in social-emotional education, this book targets feelings and the common associated facial expressions. Identifying and mimicking the facial expressions allows for awareness and social-emotional growth.

Educational Learning

The main focus of this book is MATH. Geometric Shapes, specifically. These geometric shapes used to create faces will enhance the learner's knowledge of many different measurements and angles. Introducing these shapes to your learner as early as possible will aid in retaining the information sooner. Encouraging them to point, mimick, and verbalize will produce best results.

CONFIDENT CIRCLE

CONFUSED HEPTAGON

RIGHT TRIANGLE

SAD HEART

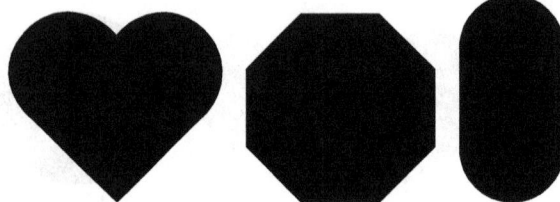

DELIGHTED NONAGON

OCTAGON

MISERABLE STAR

SUSPICIOUS DECAGON

PENTAGON

DIZZY PARALLELOGRAM

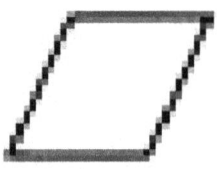

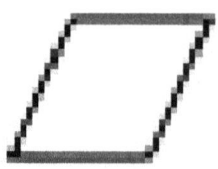

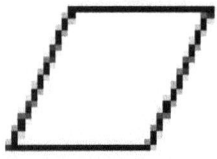

HAPPY TRAPEZOID

SQUARE

ANGRY OBTUSE TRIANGLE

WORRIED HEXAGON

DECAGON

SURPRISED RECTANGLE

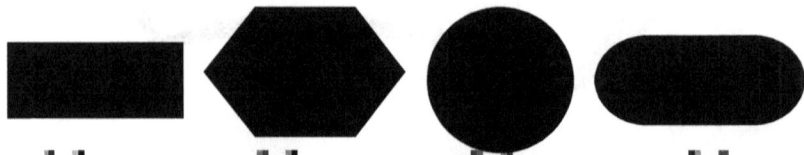

DISGUSTED   OVAL

ACUTE TRIANGLE

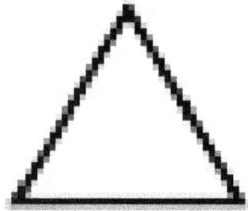

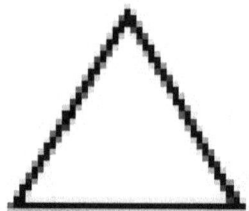

SNEAKY RHOMBUS

SLEEPY   HEXAGON

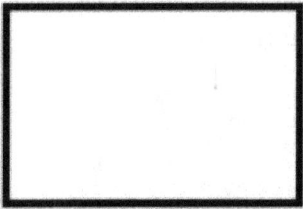
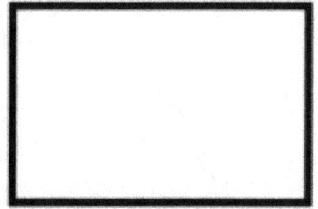

RECTANGLE

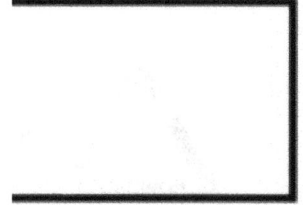

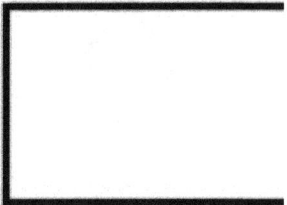
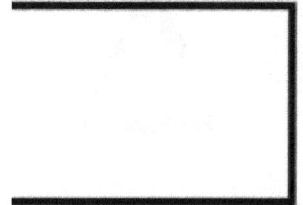

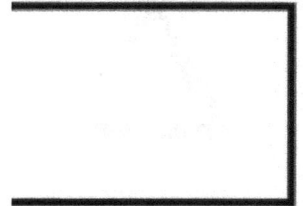

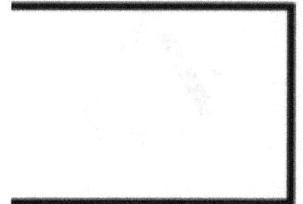

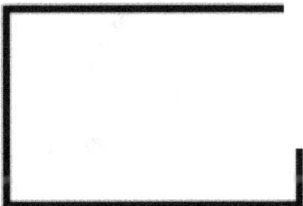
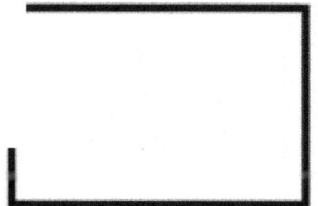

EQUILATERAL TRIANGLE

SMILING OVAL

<u>Create Your Own Face Shapes</u>

Angry Obtuse Triangle

Help the Obtuse triangle find the Angry face.

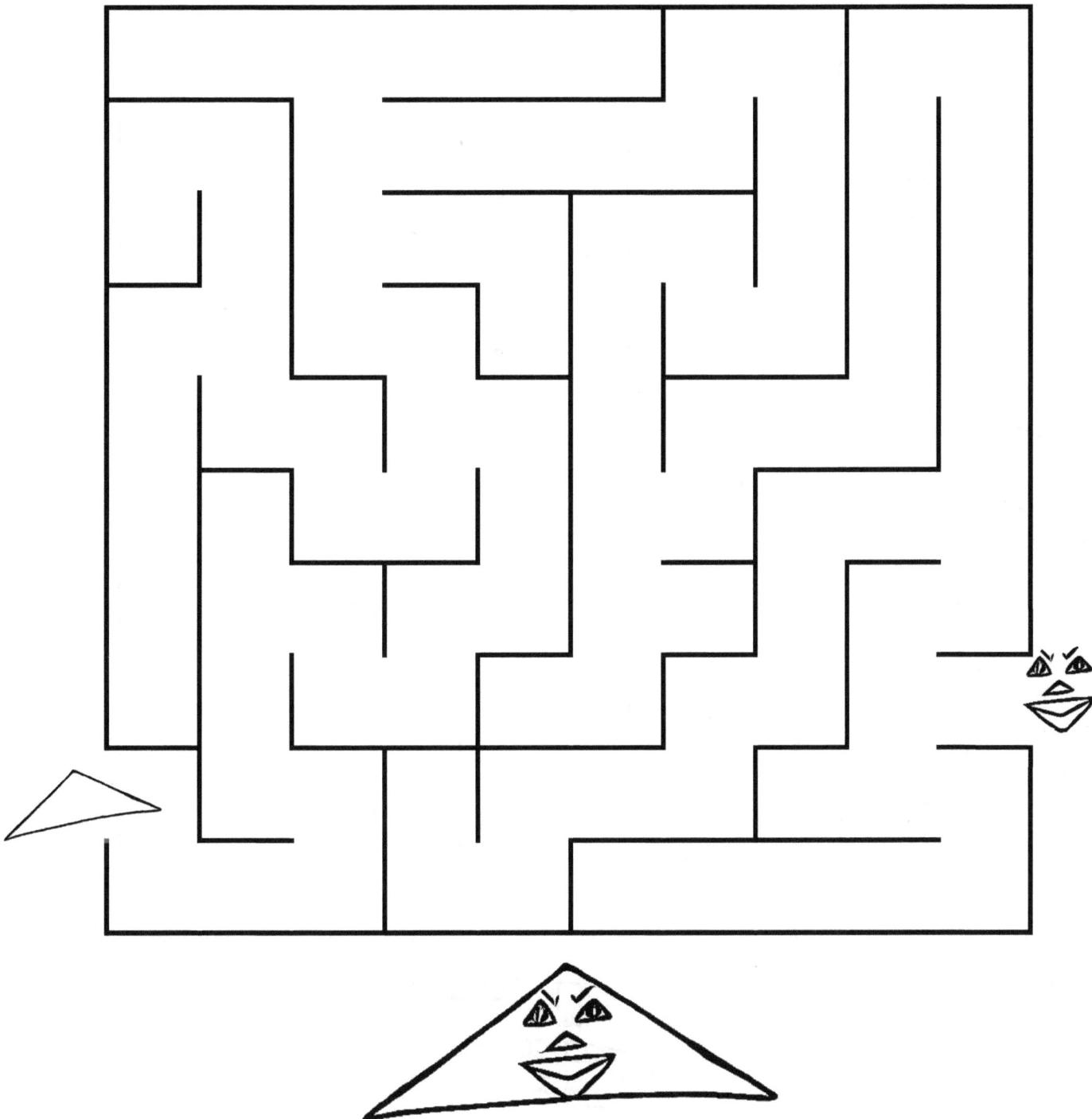

SURPRISED RECTANGLE

Help the Rectangle find the Surprised face.

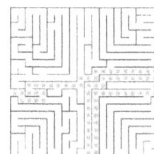

Identify the Shape

Circle the box with the correct name of the head shape.
Color the shapes.

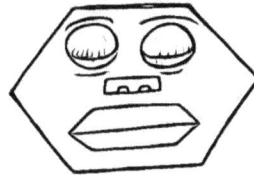

| Square | Trapezoid |
| Oval | Hexagon |

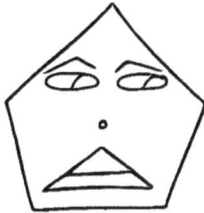

| Pentagon | Acute Triangle |
| Heptagon | Oval |

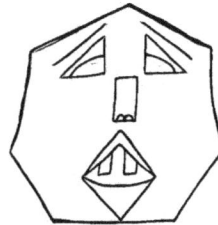

| Star | Hexagon |
| Heart | Heptagon |

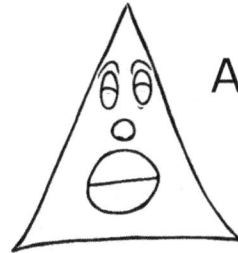

| Decagon | Heart |
| Square | Trapezoid |

| Star | Oval |
| Decagon | Acute Triangle |

EMOTIONS

WORD SEARCH

Find the words listed below and mark them.

D	A	M	I	S	E	R	A	B	L	E	S
C	O	N	F	U	S	E	D	S	A	D	U
W	B	G	G	T	U	V	S	Z	S	F	R
O	F	R	A	N	G	R	Y	T	C	D	P
R	D	I	Z	Z	Y	J	D	I	A	H	R
R	K	O	L	A	S	U	V	R	R	A	I
I	P	S	U	S	L	E	E	P	Y	P	S
E	C	O	N	F	I	D	E	N	T	P	E
D	E	L	I	G	H	T	E	D	S	Y	D

- HAPPY
- SAD
- CONFUSED
- MISERABLE
- CONFIDENT
- DIZZY
- SURPRISED
- DELIGHTED
- SCARY
- ANGRY
- WORRIED
- SLEEPY

Trace the Heptagon

COLOR THE SHAPES

HOW MANY SIDES DOES IT HAVE?

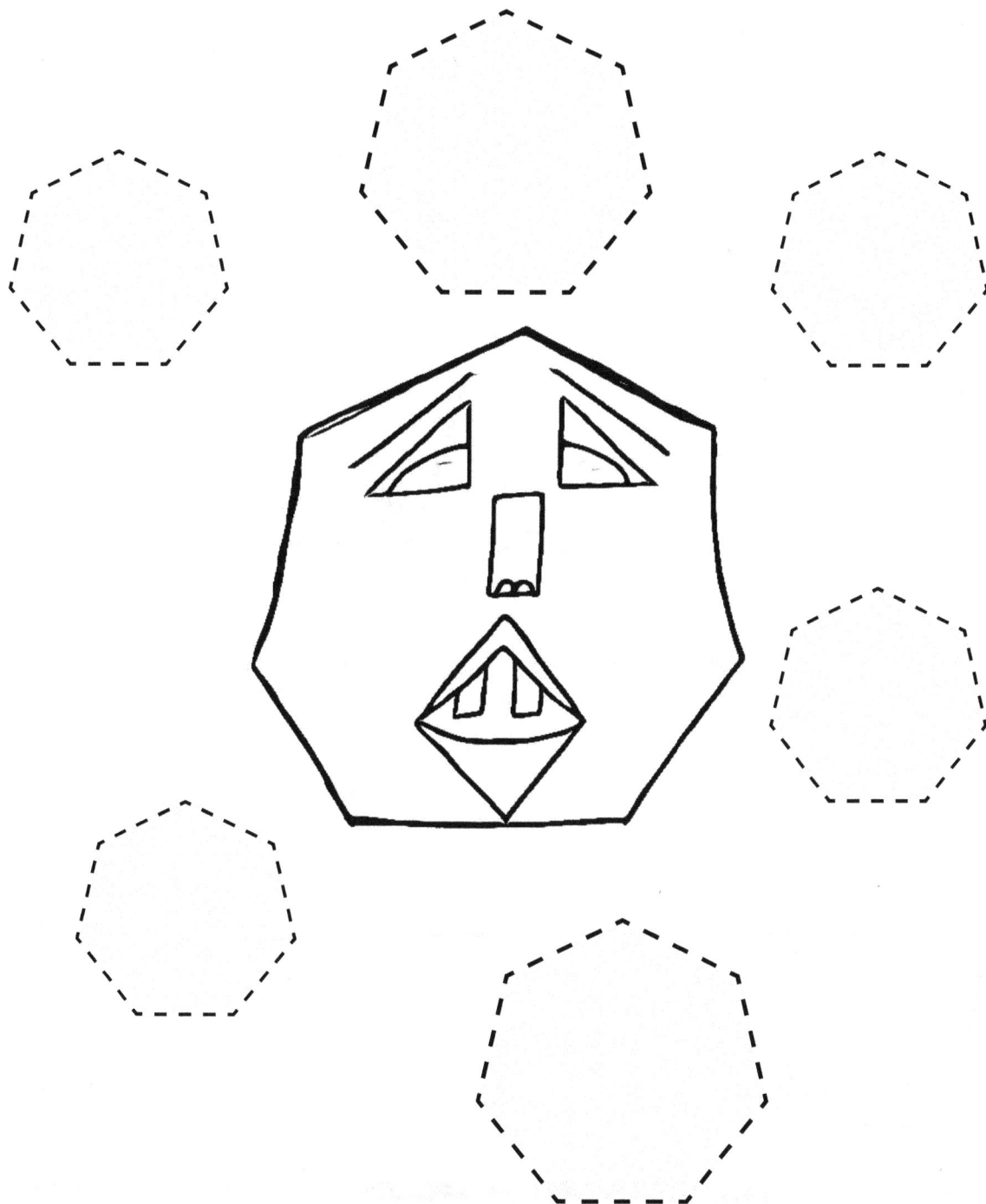

Trace the Pentagons

COLOR THE SHAPES.

HOW MANY SIDES DOES IT HAVE?

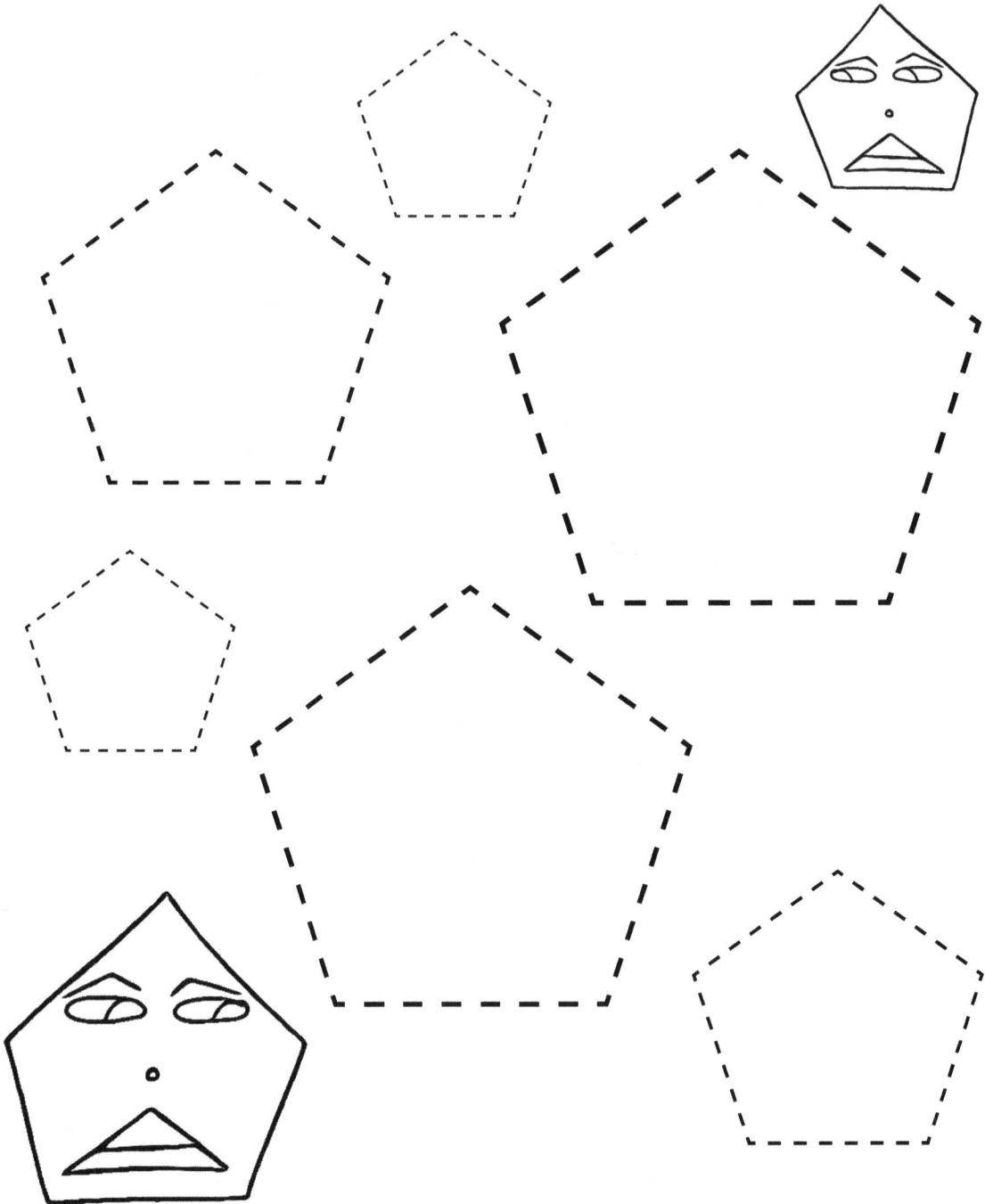

Trace the Triangles

COLOR THE SHAPES

HOW MANY SIDES DOES IT HAVE?

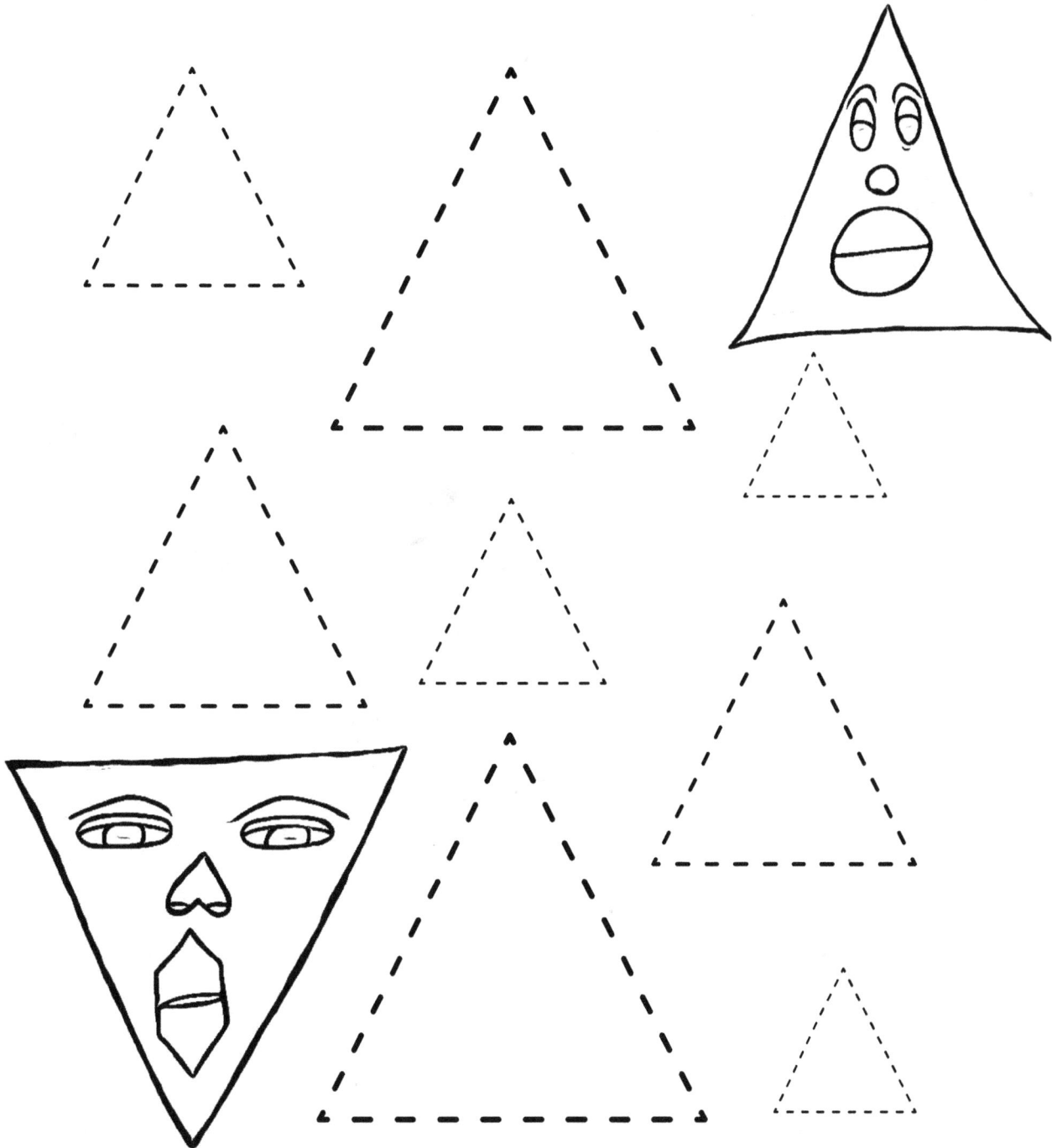

Trace the Hearts

COLOR THE SHAPES

Trace the Crescents

Color the Shapes

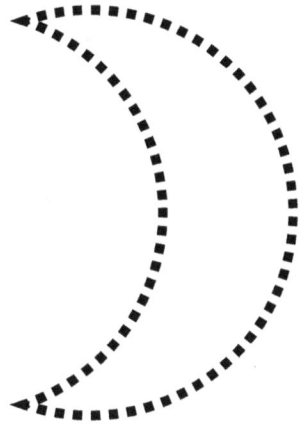

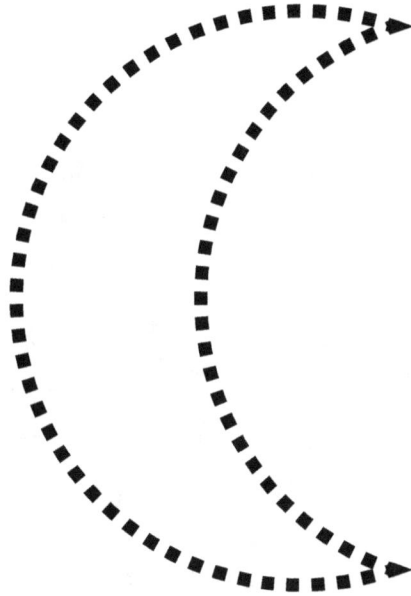

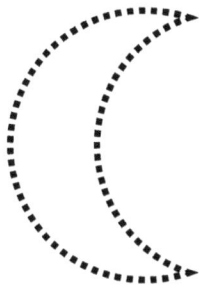

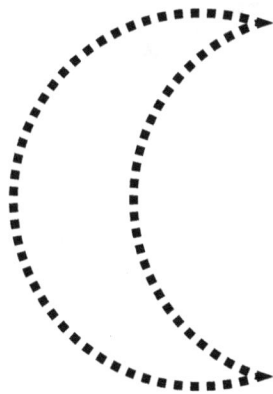

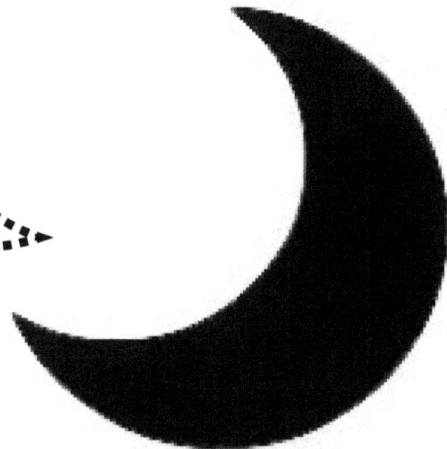

Trace the Stars

Color the Shapes

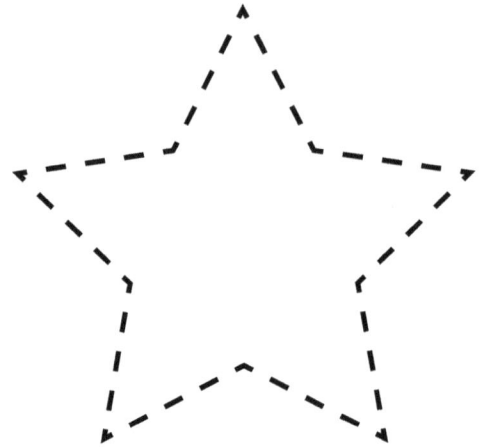

I.D. ME

Identify each shape

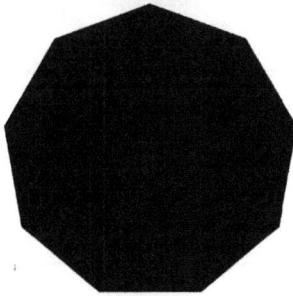

Word Categories

Write the words below into the correct boxes for polygons, ellipses , and feelings.

Triangle	Hexagon	Oval
Happy	Circle	Confident
Dizzy	Octogon	Heptagon

Polygons
_____
_____
_____
_____

Ellipses
_____
_____
_____

Feelings
_____
_____
_____

Create a sentence using two of the words above:
_____
_____

VOCABULARY WORDS

POLYGONS

QUADRILATERAL
SQUARE
RECTANGLE
PARALLELOGRAM
RHOMBUS
TRAPEZOID
TRIANGLE
STAR
PENTAGON
HEXAGON
HEPTAGON
OCTOGON
NONAGON
DECAGON

ELIPSES

OVAL
CIRCLE

SHAPES

CRESCENT
SEMI-CIRLE

EXPRESSIONS

CONFIDENT
SMILING
DISGUSTED
SNEAKY
SURPRISED
SLEEPY
WORRIED
ANGRY
HAPPY
DIZZY
MISERABLE
SUSPICIOUS
SAD
CONFUSED
DELIGHTED

Try this

Try mimicking the faces with your child. Recognize their muscular control. Can they copy the faces? Are they close to the face? Have fun!

Try drawing all your own faces on the back of each coloring page. What shapes did you use? Practice makes perfect.

Point to each element of the face and ask your child to identify the shape.

To practice math, count the total number of shapes used to create each face.

Use the 'ID ME' page to quiz your child on the geometric shapes.

Ask your child to identify the shapes of the actual faces around them. i.e. Moms face, Dads face, siblings or friends.

Encourage and Praise your child as they color. Try to withhold constructive criticism. Their talents will develop over time when they are confident.

Practice makes perfect. Color and draw as often as watching t.v. and tablets.

HAVE FUN!

DEFINITIONS

ACUTE TRIANGLE: A triangle with all the angles less than 90 degrees.

EQUILATERAL TRIANGLE: A triangle with all three sides the same length.

ISOSCELES TRIANGLE: A triangle that has two sides of equal length.

OBTUSE TRIANGLE: A triangle with angles more than 90° degrees.

RIGHT TRIANGLE: A triangle with one 90 degree angle.

PARALLELOGRAM: A rectangle with 2 acute opposite angles.

RHOMBUS: A parallelogram with opposite equal acute angles, opposite equal obtuse angles, and 4 equal sides.

SEMI-CIRCLE: A half circle.

POLYGON: A figure with 3 or more sides and angles.

QUADRILATERAL: A 4 sided figure.

PENTAGON: A figure with 5 sides and 5 angles.

HEXAGON: A figure with 6 sides and 6 angles.

HEPTAGON: A figure with 7 sides and 7 angles.

OCTAGON: A figure with 8 sides and 8 angles.

NONAGON: A figure with 9 sides and 9 angles.

DECAGON: A figure with 10 sides and 10 angles.

Disclaimer Page

Images in this book are hand drawn thus measurements and degrees may not be exact.

Each coloring page is equipt with a key code of the shapes used create that particular face.

PLEASE LEAVE US AN HONEST REVIEW.

THIS CAN BE DONE ON GOOGLE BY SCANNING THE QR CODE.

Google Review

THANK YOU

With the utmost sincerity, thank you for your purchase and participation in growing my business. There is no AHotMess without YOU!

AHotMessDotCom

Enjoy your purchase?

Check out the 'I Can Draw' Sketch Book. This is not your average sketch book. See for yourself!

Also read the 'BUT WHAT AM I FEELING?' story book. A rhyming book about social - emotional awareness and neurodivergence.

SCAN HERE

Finally, lets keep in touch. Stay in the know of what going on with AHotMessDotCom & AHotMessDotEDU

www.ingramcontent.com/pod-product-compliance
Lightning Source LLC
Chambersburg PA
CBHW062109090426

42741CB00015B/3374